COSMIC CHRONICLES

Neil deGrasse Tyson and the Life Cycle of Stars

Oteren.Fredrick

DISCLAIMER

COPYRIGHT

TABLE OF CONTENT

INTRODUCTION

The universe is a fabulous embroidery of steadily changing heavenly peculiarities, and among its most captivating components are stars. These brilliant circles of hot plasma enlighten the night sky as well as assume a significant part in the enormous pattern of issue and energy. Their lives, from birth to death, shape the actual texture of the universe.

In this book, Cosmic Narratives: Neil deGrasse Tyson and the Existence Pattern of Stars, we set out on an excursion through the heavenly life cycle, directed by the experiences of one of the most conspicuous astrophysicists within recent memory, Neil deGrasse Tyson. Tyson's commitments to how we might interpret the universe are both significant and available, making him an optimal focal point through which to investigate the intricate and dynamic course of heavenly advancement.

This presentation makes way for our investigation of how stars structure, advance, and eventually meet their end. We will dive into the primary standards of heavenly development, laying the preparation for understanding how these divine articles impact the universe. We will likewise give a concise outline of Neil deGrasse Tyson's vocation and his effect on the field of astronomy, featuring his capacity to discuss complex logical ideas with lucidity and energy.

Toward the finish of this book, perusers will acquire a more profound appreciation for the existence pattern of stars and the huge job they play in the vast dance of creation and obliteration. Through Tyson's points of view and speculations, we will uncover the secrets of heavenly advancement and its suggestions for the universe in general.

CHAPTER 1:

The Birth of Stars

The universe, in the entirety of its wonder and intricacy, is a huge span where stars are conceived, live, and pass on in a nonstop pattern of vast change. The introduction of a star is quite possibly of the most captivating and basic cycle in astronomy, making way for the heap peculiarities that follow. In this part, we will investigate the multifaceted cycles that lead to star development, directed by the experiences of Neil deGrasse Tyson, whose clarifications have enlightened how we might interpret these divine monsters.

Heavenly Nurseries: Nebulae and Sub-atomic Clouds

Stars don't just show up out of the void; their beginnings are established in the tremendous, thick billows of gas and residue dissipated all through the universe. These mists, known as nebulae, are frequently alluded to as heavenly nurseries since they act as the essential destinations for star development. The most popular illustration of a cloud is the Orion Cloud, a wonderful and energetic district where new stars are constantly being conceived.

Nebulae are made dominatingly out of hydrogen, the easiest and most bountiful component in the universe, alongside more modest measures of helium and minor components. These mists can be monstrous, containing a few times the mass of the Sun and crossing light-years across. The thickness and temperature of the gas inside a cloud are for the most part low, however locales of higher thickness, known as sub-atomic mists, give the circumstances important to star development.

Neil deGrasse Tyson frequently portrays sub-atomic mists as the "inestimable favorable places" for stars. These mists are so cold, with temperatures decreasing to only two or three many degrees above outright zero, that the gas and residue inside them can bunch together under their own gravity. As these clusters develop denser, they start to implode internal, making locales of expanding tension and temperature. This cycle is the underlying move toward star arrangement.

The Course of Star Formation

The progress from a chilly, diffuse cloud to a hot, iridescent star includes a few key stages. The primary stage, known as the protostar phase, starts when a cluster of gas and residue inside a sub-atomic cloud turns out to be adequately thick to begin falling under its own gravity. As the material falls internal, it frames a focal center that becomes progressively hot and thick. This center is encircled by an alternating plate of gas and residue, and the whole design is known as a protostar.

During the protostar stage, the focal center keeps on accumulating material from the encompassing plate. As the center's temperature climbs, atomic combination responses in the end start, denoting the introduction of another star. In any case, before these responses can support themselves, the protostar should arrive at a minimum amount and temperature. Neil deGrasse Tyson makes sense of that the start of atomic combination is similar to lighting up a match; when the interaction starts, the star enters another period of its life cycle.

Tyson's Bits of knowledge on Protostars and Accretion

Neil deGrasse Tyson gives a convincing clarification of the growth cycle, which is vital for the development of a star. In his public talks and works, Tyson frequently compares the growth circle to a turning pizza batter, where material from the encompassing cloud falls internal and collects around the shaping star. This similarity outlines the unique idea of the growth cycle and the job it plays in molding the early star.

The material in the gradual addition plate adds to the star's development as well as impacts its turn and attractive field. As the protostar keeps on social event mass, it creates strong heavenly breezes that can blow away overabundance gas and residue. These breezes assume a vital part in clearing the encompassing district and permitting the recently shaped star to completely arise.

Tyson's bits of knowledge stretch out to the changeability in star arrangement rates and results. He underlines that not all stars are brought into the world with similar attributes; some structure rapidly and become monstrous stars, while others develop all the more leisurely and stay more modest. This changeability is a consequence of contrasts in the underlying states of the sub-atomic cloud and the productivity of the growth cycle.

The Job of Attractive Fields and Jets

Notwithstanding gravity, attractive fields assume a critical part during the time spent star development. Attractive fields can impact the elements of the gradual addition plate and the progression of material onto the protostar. Tyson features the significance of these fields in molding the last attributes of the star, including its revolution rate and attractive action.

One more entrancing part of star development is the presence of jets — slender surges of charged particles catapulted from the shafts of the protostar. These planes are a consequence of the perplexing communications between the protostar's attractive field and the growth circle. As material falls onto the protostar, some of it is diverted along the attractive field lines and removed at high speeds. Tyson depicts these planes as the "infinite input instrument" that controls the star's development and the general climate.

Contextual analysis: The Orion Nebula

To show the cycles talked about, let us think about the Orion Cloud, one of the most all around concentrated on districts of star arrangement. Found roughly 1,300 light-years from Earth, the Orion Cloud is a dynamic and dynamic heavenly nursery where many youthful stars are being conceived.

The Orion Cloud is a perfect representation of how sub-atomic mists breakdown to shape new stars. Perceptions with telescopes, for example, the Hubble Space Telescope have given point by point pictures of the cloud's multifaceted design, uncovering the thick centers of gas and residue where star development is occurring. Tyson frequently references the Orion Cloud in his conversations to underline the scale and excellence of heavenly birth.

In the core of the Orion Cloud, cosmologists have distinguished various protostars and youthful heavenly items, each in various transformative phases. These perceptions give important bits of knowledge into the cycles that administer star arrangement and assist with refining how we might interpret the components in question.

CHAPTER 2:

Main Sequence Stars

The essential gathering is a significant stage in the presence example of a star, meaning a period of safety and predictable energy yield. During this stage, stars go through nuclear mix at their focuses, changing over hydrogen into helium and making the energy that upholds their sparkle and power. This part plunges into the traits of chief gathering stars, their life cycles, and Neil deGrasse Tyson's encounters into their approach to acting and significance.

Sorting out Essential Game plan Stars

The essential gathering is where a star consumes the majority of its time on earth. For our Sun, this stage addresses around 90% of its finished future. During this period, a star wires hydrogen particles in its middle, making helium and conveying a monster proportion of energy all the while. This energy sends outward, adjusting the force of gravity and staying aware of the star's agreement.

The imperative part of essential progression stars is their hydrostatic equilibrium — a touchy congruity between the interior draw of gravity and the outward pressure from the energy made by nuclear blend. This balance ensures that the star stays stable, with its outside layers neither collapsing inward nor broadening outward fiercely.

Neil deGrasse Tyson regularly highlights the meaning of this equilibrium in his explanations. He looks at the connection to a particularly changed radiator: the power created by mix gives the essential strain to really look at gravitational breakdown, making a reliable, stable star.

The Occupation of Nuclear Fusion

At the center of a chief gathering star is the course of nuclear mix. In the middle, temperatures show up at countless degrees Celsius, and the strain is huge. These conditions are great for nuclear reactions to occur, where hydrogen centers (protons) unite to shape helium centers. This cycle conveys a great deal of energy as light and force.

Tyson gives an unmistakable similitude to getting a handle on this cycle: he takes a gander at the blend reactions in a star's middle to the consuming of fuel in an engine. Correspondingly as consuming fuel produces energy to drive a vehicle, nuclear mix produces energy to fuel a star's sparkle and power. In any case, unlike an engine, the blend cycle in stars is upheld more than billions of years in view of the gigantic pressures and temperatures included.

The speed of mix reactions and, in this way, the star's radiance and not totally settled by its mass. More tremendous stars have higher focus temperatures and strains, inciting speedier mix rates and more conspicuous energy yield. Of course, less gigantic stars have cooler focuses and all the more sluggish mix rates.

Tyson's Perspective on Brilliant Equilibrium

Neil deGrasse Tyson's discussions on brilliant congruity highlight the delicate balance expected to keep a consistent star. He figures out that this harmony is basic for a star's life expectancy and strength. Without the genuine concordance, a star could either fall under its own gravity or develop fiercely.

In his public discussions, Tyson habitually uses the comparability of an inflatable to outline this harmony. He portrays how the inflatable's surface strain and the external vaporous pressure ought to be acclimated to keep the inflatable stable. In like manner, in a star, the inside pressure from mix reactions ought to check the gravitational powers endeavoring to pull the star inside.

Tyson furthermore addresses how different stars achieve this harmony considering their mass. Tremendous stars have more grounded gravitational abilities and, along these lines, require higher obliges from mix reactions to stay aware of safety. This results in more restricted futures anyway more outrageous energy yields. Curiously, more unobtrusive stars have more weak gravitational abilities and can stay aware of dauntlessness with lower mix rates, provoking longer futures.

The Hertzsprung-Russell Diagram

The Hertzsprung-Russell (H-R) diagram is a basic gadget for understanding the different periods of a star's life cycle, including the essential gathering. This diagram plots stars as demonstrated by their brilliance (wonder) and surface temperature. Essential progression stars are found along a slanting band that connects from the upper left (hot, brilliant stars) to the lower right (cool, faint stars).

The put of a star on the H-R outline gives critical information about its characteristics and stage in the magnificent life cycle. For example, the Sun is arranged in the important progression band, exhibiting its normal luminosity and temperature stood out from various stars.

Tyson much of the time implies the H-R chart to figure out the assortment among head progression stars. He points out that while all essential progression stars share the typical nature of hydrogen mix, they change commonly in their radiance, temperature, and size. This assortment is a result of the different masses and manifestations of stars.

The Presences of Essential Progression Stars

The length of a star's insight on the crucial gathering depends upon its mass. Tremendous stars, for instance, those on different occasions the mass of the Sun, consume their hydrogen fuel even more rapidly. Consequently, they invest a decently short energy outline in the essential progression stage — several million years — before progressing to the accompanying periods of their life cycle.

Strangely, more unobtrusive stars, similar to red small individuals, have a ton of lower mix rates and in this manner can remain on the chief gathering for tens to a long time. These stars have much more sluggish speed of hydrogen consuming and can stay aware of their unfaltering quality for far longer than their more gigantic accomplices.

Tyson includes the consequences of these differentiations in glorious futures. He gets a handle on that while massive stars could grow quickly and end their lives in profound grandiose blast impacts, more unassuming stars like red small individuals have an all the more consistent and long turn of events. This really long stage grants red dwarfs to contribute every time to the substance improvement of the grandiose framework over an extensive period.

Logical examination: The Sun

To portray the characteristics of essential gathering stars, let us consider our Sun, an extraordinary portrayal of a medium-sized basic progression star. The Sun has been in the essential gathering stage for generally 4.6 billion years and should remain in this stage for another 5 billion years preceding forming into a red goliath.

The Sun's sufficiency and brilliance are an outcome of the ideal congruity between gravitational powers and the energy conveyed by nuclear mix. Tyson habitually includes the Sun as a benchmark for figuring out sublime cycles and characteristics, highlighting its work in supporting life in the world and its impact on the close by planet bunch.

CHAPTER 3:

The Lives of Stars: From Red Dwarfs to Giants

The existence of a star is a unique excursion set apart by phases of change and development. Subsequent to spending most of their lives on the fundamental arrangement, stars go through tremendous changes relying upon their mass. This part investigates the various ways that stars take as they develop past the fundamental succession, zeroing in on the change from red midgets to goliaths, and Neil deGrasse Tyson's experiences into these phases of heavenly advancement.

Attributes of Red Smaller people and Their Longevity

Red diminutive people are among the most widely recognized sorts of stars known to man. They are little, cool, and faint contrasted with their more gigantic partners. With masses going from around 0.08 to 0.5 times that of the Sun, red midgets consume their hydrogen fuel gradually, permitting them to stay on the fundamental succession for remarkably significant stretches — frequently up to 100 billion years or more.

One of the vital elements of red midgets is their low iridescence and surface temperature. They radiate considerably less light contrasted with bigger stars and have temperatures going from 2,500 to 4,000 Kelvin. Regardless of their unobtrusive appearance, red diminutive people assume a urgent part in the cosmic environment, adding to the substance improvement of the universe over their drawn out life expectancies.

Neil deGrasse Tyson underlines the significance of red diminutive people in his conversations on heavenly advancement. He brings up that these stars are the most various in the cosmic system and their drawn out dependability permits them to give a steady wellspring of energy north of billions of years. Tyson's work additionally features the possible livability of planets circling red smaller people, as their long life expectancies could give more than adequate opportunity to life to create.

The Development of Halfway Mass Stars

Moderate mass stars, with masses going from around 0.5 to multiple times that of the Sun, follow an alternate developmental way contrasted with red smaller people. These stars experience a more sensational change once they exhaust their hydrogen fuel.

As transitional mass stars approach the finish of their principal arrangement stage, they start to go through changes in their center and external layers. The center agreements and warms up while the external layers grow and cool, changing the star into a red goliath. This stage is described by a huge expansion in iridescence and an emotional extension of the star's external envelope.

Neil deGrasse Tyson frequently involves the red goliath stage as an illustration to delineate the emotional changes that stars go through. He depicts the extension of a red goliath as a "inestimable inflatable" that grows to ordinarily its unique size. This extension is a consequence of the expanded energy creation in the center and the complicated transaction of different powers inside the star.

The Development of Planetary Nebulae and White Dwarfs

At the point when halfway mass stars arrive at the finish of their red monster stage, they shed their external layers into space, making a planetary nebula. This cloud is a shining shell of gas removed from the star's external envelope. The center abandoned, as of now not upheld by combination responses, agreements and cools to turn into a white dwarf.

White midgets are thick, hot leftovers of stars that have depleted their atomic fuel. They have masses practically identical to the Sun however are just about the size of Earth. Over the long haul, white diminutive people cool and blur, at last becoming black dwarfs, however this cycle takes more time than the ongoing age of the universe.

Tyson's clarifications frequently stress the meaning of planetary nebulae and white diminutive people in the heavenly lifecycle. He takes note of that the material ousted during the red monster stage improves the interstellar medium with weighty components, adding to the development of new stars and planetary frameworks

Gigantic Stars and Their Transformation

Gigantic stars, with masses more prominent than multiple times that of the Sun, follow an alternate transformative track. These stars go through additional perplexing cycles and experience more limited life expectancies contrasted with lower-mass stars. Subsequent to leaving the fundamental succession, huge stars venture into supergiants, with their external layers becoming essentially bigger than those of red goliaths.

The center of a huge star agreements and warms up, prompting a progression of combination responses that produce heavier components like carbon, oxygen, and iron. This stage is set apart by extraordinary energy yield and an exceptionally powerful climate. The external layers of the star become unsteady and are at last removed in an emotional supernova blast.

Neil deGrasse Tyson frequently talks about supernovae as inestimable occasions vital. He makes sense of that supernovae are answerable for scattering weighty components all through the universe, adding to the arrangement of new stars and planetary frameworks. The blast of an enormous star likewise abandons a minimal remainder, which can be a neutron star or a dark opening, contingent upon the mass of the first star.

Neutron Stars and Dark Holes

The leftovers of supernovae can bring about two sorts of heavenly remainders: neutron stars and black holes. Neutron stars are extraordinarily thick articles made basically out of neutrons, with masses going from around 1.4 to 2.16 times that of the Sun, however with radii of something like 10 kilometers. They are the remainders of stars that were not adequately huge to frame dark openings.

Dark openings, then again, are locales of spacetime with gravitational powers serious areas of strength for so not even light can get away. They are shaped when the center of a monstrous star implodes under its own gravity to a place of boundless thickness. The limit encompassing a dark opening, known as the occasion skyline, denotes the point past which no data can escape.

Tyson regularly addresses the puzzling idea of dark openings in his conversations. He features the continuous examination and revelations connected with these items, remembering their job for cosmic elements and their possible associations with other central parts of physical science.

Contextual analysis: Betelgeuse and Sirius

To represent the assorted ways of heavenly advancement, let us analyze two notable stars: Betelgeuse and Sirius. Betelgeuse, situated in the star grouping Orion, is a red supergiant that is approaching a mind-blowing finish. It is an illustration of a huge star going through the red monster stage and is supposed to detonate as a cosmic explosion ultimately.

Sirius, then again, is a parallel star framework comprising of Sirius A, a principal succession star, and Sirius B, a white smaller person. Sirius A will be a moderately youthful and hot star, while Sirius B addresses the end phase of a middle of the road mass star's life. The difference between these two stars features the different developmental ways taken by stars of shifting masses.

CHAPTER 4:

<u>The Death of Stars</u>

The demise of a star is an emotional and complex interaction that connotes the finish of its dynamic life cycle. Contingent upon the star's mass and structure, this last stage can bring about different divine peculiarities, including white midgets, neutron stars, or dark openings. In this part, we will investigate the various situations for the demise of stars, looking at the jobs of supernovae, heavenly leftovers, and the bits of knowledge given by Neil deGrasse Tyson.

<u>White Diminutive people: The End Phase of Low-Mass Stars</u>

For stars with masses up to multiple times that of the Sun, the finish of their life cycle comes full circle in the development of a white dwarf. A white diminutive person is the thick, hot remainder abandoned after the external layers of a star have been shed. This cycle starts when a low-to moderate mass star debilitates its atomic fuel, stopping combination responses in its center.

As the star leaves the fundamental succession and enters the red goliath stage, it sheds its external layers into space, making a planetary nebula. The center that remains is incredibly warm and thick, with a mass tantamount to that of the Sun however a sweep like that of Earth. Over the long run, the white bantam cools and blurs, ultimately turning into a black dwarf, however the universe isn't mature enough for any dark diminutive people to exist yet.

Neil deGrasse Tyson frequently portrays white diminutive people as the "enormous remainders" of heavenly advancement, underlining their job in the reusing of materials inside the world. He features the meaning of white midgets in adding to the compound improvement of the universe, as the material removed during the red goliath stage is reallocated and can frame new stars and planetary frameworks.

Neutron Stars: The Leftovers of Supernovae

More enormous stars, with beginning masses more noteworthy than multiple times that of the Sun, go through an alternate finish of-life situation. At the point when these stars exhaust their atomic fuel, their centers breakdown under gigantic gravity, prompting a supernova blast. The external layers of the star are ousted into space, while the center remaining parts and turns into a neutron star.

Neutron stars are inconceivably thick items made for the most part out of neutrons. They have masses between around 1.4 and 2.16 times that of the Sun yet are something like 10 to 20 kilometers in breadth. The thickness of a neutron star is outrageous to the point that a sugar-block measured measure of neutron-star material would weigh around 100 million tons on The planet.

Tyson makes sense of the exceptional properties of neutron stars in his conversations on heavenly leftovers. He underlines their job in astrophysical peculiarities, like pulsars — quickly pivoting neutron stars that discharge light emissions noticeable from Earth. Neutron stars likewise assume a critical part in grasping the constraints of atomic matter and the way of behaving of issue under outrageous circumstances.

Dark Openings: A definitive End State

For the most huge stars, the finish of their life cycle can bring about the development of a black hole. Whenever an enormous star's center falls under its own gravity, it can pack to where the break speed surpasses the speed of light, making a locale of spacetime with an occasion skyline past which nothing can escape. This is the central trait of a dark opening.

Dark openings come in different sizes, from heavenly mass dark openings framed by the breakdown of individual stars to supermassive dark openings found at the focuses of systems. The course of dark opening development includes complex elements, including the cooperation of gravitational waves and the potential for consolidations with other dark openings.

Neil deGrasse Tyson frequently addresses dark openings in his conversations, featuring their baffling nature and the continuous examination into their properties. He makes sense of how dark openings impact their environmental factors, for example, accumulating matter from friend stars or taking care of the development of universes through their gravitational impact.

Supernovae: Enormous Blasts and Synthetic Enrichment

Supernovae are among the most awesome occasions known to mankind. These enormous blasts happen toward the finish of a star's life when the center breakdowns, and the external layers are catapulted into space. The blast delivers an enormous measure of energy, eclipsing whole worlds for a brief period and dispersing components into the interstellar medium.

Neil deGrasse Tyson stresses the job of supernovae in grandiose science. He makes sense of that these blasts are answerable for scattering weighty components like iron, nickel, and different metals all through the system. These components are fundamental for the arrangement of planets and life, featuring the interconnected idea of heavenly and planetary cycles.

Supernovae likewise assume a critical part in the development of neutron stars and dark openings. The elements of the blast and the breakdown of the center outcome in these thick leftovers, which impact the development of the encompassing world. Tyson's experiences into supernovae highlight their importance in both heavenly advancement and vast science.

Contextual investigation: SN 1987A

One of the most renowned supernovae in ongoing history is SN 1987A, which happened in the Huge Magellanic Cloud, a close by world. This cosmic explosion gave significant bits of knowledge into the cycles of heavenly demise and the development of neutron stars. Perceptions of SN 1987A uncovered insights concerning the blast elements, the development of a neutron star, and the dispersal of weighty components into space.

Tyson habitually references SN 1987A to outline the down to earth ramifications of cosmic explosion research. He features how the investigation of this occasion has progressed how we might interpret cosmic explosion mechanics and the job of these blasts in enhancing the interstellar medium with fundamental components.

The Effect of Heavenly Demise on Cosmic Evolution

The demise of stars has significant ramifications for the advancement of systems. The material shot out by kicking the bucket stars adds to the arrangement of new stars and planetary frameworks, impacting the substance piece and design of worlds. The cycles of supernovae, white bantam development, and dark opening creation all assume a part in molding the cosmic scene.

Tyson's work frequently investigates the more extensive effect of heavenly demise on universe development and advancement. He examines how the reusing of heavenly materials and the impact of heavenly remainders add to the continuous cycles of star arrangement and cosmic elements. The transaction between heavenly life and passing is a critical figure the development of the universe.

CHAPTER 5:

The Role of Stellar Evolution in Galactic Dynamics

Heavenly development isn't just significant for understanding the existence patterns of individual stars yet additionally for fathoming the more extensive elements of systems. The cycles of star arrangement, development, and demise fundamentally influence the design, sythesis, and conduct of universes. In this section, we will investigate what heavenly advancement means for cosmic elements, zeroing in on the jobs of supernovae, heavenly criticism, and the arrangement of new stars, with experiences from Neil deGrasse Tyson.

Heavenly Input and Cosmic Ecosystems

Heavenly criticism alludes to the different manners by which stars impact their environmental factors all through their life cycles. This criticism happens through various systems, including heavenly breezes, cosmic explosion blasts, and the radiation produced by stars. These cycles assume a vital part in molding the interstellar medium and impacting star development inside systems.

1. Stellar Winds:

As stars advance, especially during their red monster or supergiant stages, they produce strong heavenly breezes — surges of charged particles that blow away material from their external layers. These breezes add to the advancement of the interstellar medium with weighty components and influence the thickness and dissemination of gas inside the system.

Neil deGrasse Tyson frequently underscores the significance of heavenly breezes in forming cosmic conditions. He makes sense of how these breezes can make cavities or air pockets in the interstellar medium, affecting the development of new stars and changing the design of worlds.

2. Supernova Explosions:
Supernovae are among the most lively occasions known to man. At the point when monstrous stars detonate, they discharge tremendous measures of energy and dissipate components all through the universe. This material improves the interstellar medium with weighty components, which are fundamental for

the development of new stars and planetary frameworks.

Tyson examines how supernovae add to the compound improvement of worlds and drive the elements of star arrangement. He features that the shock waves from supernovae can set off the breakdown of neighboring gas mists, prompting the development of new stars and impacting the general star arrangement rate inside a cosmic system.

The Development of New Stars and Heavenly Nurseries

The cycles of heavenly criticism and the reusing of material from passing on stars assume a vital part in the development of new stars. The interstellar medium, advanced with weighty components from past ages of stars, fills in as the natural substance for new star development.

1. Molecular Clouds:

Sub-atomic mists are thick locales of gas and residue where new stars are conceived. These mists frequently structure because of the pressure of interstellar gas by shock waves from supernovae or cooperations with heavenly breezes. The breakdown of these mists under gravity prompts the arrangement of protostars and in the end new stars.

Tyson frequently alludes to sub-atomic mists as "heavenly nurseries" where the up and coming age of stars and planetary frameworks is shaped. He underscores the repetitive idea of star arrangement, where the material ousted by passing on stars gives the structure blocks to future stars.

2. Star Clusters:
New stars frequently structure in bunches, where numerous stars are brought into the world from a similar sub-atomic cloud. These bunches can go from little gatherings to huge affiliations containing large number of stars. The elements of star groups, including their collaborations and

the dispersal of their individuals, assume a part in the development of the cosmic climate.

Tyson talks about the meaning of star groups in grasping heavenly populaces and cosmic elements. He makes sense of how the investigation of star bunches gives bits of knowledge into the states of the early universe and the cycles of star development.

Cosmic Elements and Structure

The impact of heavenly advancement reaches out to the huge scope design and elements of systems. The circulation and conduct of stars, alongside the impacts of heavenly criticism, shape the general design of cosmic systems.

1. Galaxy Development and Evolution:
The course of universe arrangement includes the collection of gas, residue, and stars north of billions of years. Heavenly development assumes a part in molding the construction of cosmic systems through cycles like the

collection of weighty components, the arrangement of various heavenly populaces, and the dissemination of mass.

Tyson features the job of heavenly development in cosmic system arrangement, stressing how the burnning of material through heavenly life cycles adds to the arrangement of various kinds of worlds. He examines how systems develop over the long run, affected by the continuous cycles of star arrangement, heavenly input, and cooperations with different universes.

2. Stellar Populaces and Cosmic Components:
Systems are made out of different parts, including the plate, lump, and corona. Every part has an alternate heavenly populace, reflecting various phases of heavenly development. For instance, the plate of a cosmic system contains more youthful, fundamental succession stars, while the radiance contains more seasoned, more developed stars.

Tyson frequently examines the meaning of concentrating on various heavenly populaces in figuring out cosmic design. He makes sense of how the circulation of stars and their properties give bits of knowledge into the set of experiences and elements of systems.

Contextual analysis: The Smooth Way Galaxy

To represent the effect of heavenly advancement on cosmic elements, let us inspect our own universe, the Milky Way. The Smooth Way is a banished winding world with a perplexing design, including a focal lump, a plate with twisting arms, and an encompassing radiance. The circulation of stars and the presence of various heavenly populaces mirror the cycles of heavenly development and criticism.

Neil deGrasse Tyson much of the time involves the Smooth Way as a contextual investigation to make sense of the impact of heavenly development on cosmic elements. He examines how the arrangement of various parts of the world is connected with the cycles of star development, heavenly input, and the pushing of material through heavenly life cycles.

CHAPTER 6:

The Universe at Large: Cosmic Evolution and the Interplay of Forces

The life and demise of stars are fundamental to the fantastic story of the universe, impacting infinite development for an immense scope. From the arrangement of worlds to the appropriation of components across the universe, heavenly cycles shape the construction and advancement of the universe. This section investigates the more extensive setting of inestimable development, zeroing in on the exchange of powers that drive the advancement of the universe and the job of stars inside this stupendous system. Bits of knowledge from Neil deGrasse Tyson will enlighten how these cycles associate with the more extensive comprehension of inestimable peculiarities.

Vast Development and the Arrangement of Structure

The universe has developed fundamentally since the Enormous detonation, bringing about the arrangement of worlds, stars, and planetary frameworks. Heavenly development assumes a vital part in this cycle, impacting the conveyance of components and the development of enormous designs.

1. Big Bang Nucleosynthesis:
The universe started with the Huge explosion, an occasion that created the principal components, essentially hydrogen and helium.

This early stage nucleosynthesis set up for later heavenly cycles. As the universe extended and cooled, districts of gas started to implode under gravity, framing the main stars and systems.

Tyson accentuates the meaning of the early universe in his conversations on vast development. He makes sense of how the underlying circumstances and essential overflows laid out during the Huge explosion impacted the ensuing development of stars and cosmic systems.

2. Galaxy Arrangement and Evolution:
Worlds framed from the breakdown of gas mists and the conglomeration of dull matter. Heavenly development inside cosmic systems adds to their construction and creation, with stars assuming a urgent part in improving the interstellar medium and driving the elements of universe arrangement.

Tyson frequently talks about how cosmic systems develop after some time, impacted by interior cycles, for example, star arrangement and outside factors like connections with different worlds. He features how the investigation of world development and advancement gives bits of knowledge into the more extensive history of the universe.

The Job of Dim Matter and Dull Energy

The universe is administered by powers past the noticeable matter of stars and cosmic systems. Dull matter and dim energy are pivotal parts that impact the design and development of the universe.

1. Dark Matter:
 Dim matter is an undetectable type of issue that doesn't radiate light or connect with electromagnetic powers. Its presence is deduced from its gravitational impacts on apparent matter, for example, the turn bends of systems and the movement of world bunches. Dim matter

assumes a critical part in the development and dependability of grandiose designs.

Tyson examines the significance of dull matter in grasping the huge scope design of the universe. He makes sense of how dull matter impacts the dissemination of systems and the elements of vast development, in spite of being imperceptible through customary means.

2. Dark Energy:
Dull energy is a baffling power driving the sped up extension of the universe. It balances the impacts of gravity, making the universe grow at a rising rate. The idea of dim energy stays quite possibly of the greatest inquiry in cosmology.

Tyson frequently investigates the ramifications of dim energy for the eventual fate of the universe. He features how the revelation of dim energy has changed how we might interpret inestimable advancement and a definitive destiny of the universe.

The Effect of Vast Powers on Heavenly and Cosmic Processes

Enormous powers, for example, gravity, radiation pressure, and attractive fields assume a pivotal part in molding the development of stars and cosmic systems. These powers impact different parts of heavenly life cycles and cosmic elements.

1. Gravity and Star Formation:
Gravity is the essential power driving the breakdown of gas mists and the arrangement of new stars. The exchange of gravitational powers inside atomic mists prompts the arrangement of protostars and the inception of atomic combination.

Tyson much of the time examines how gravity administers the arrangement and advancement of inestimable designs. He makes sense of how gravitational cooperations shape the introduction of stars and the improvement of cosmic systems.

2. Radiation Strain and Heavenly Evolution:
Radiation strain from gigantic stars influences their advancement and the encompassing interstellar medium. The extreme radiation transmitted by high-mass stars can impact the development of new stars and the design of sub-atomic mists.

Tyson features the job of radiation strain in his conversations on heavenly advancement. He makes sense of what the energy result of stars means for their environmental elements and adds to the elements of star development.

3. Magnetic Fields and Heavenly Activity:
Attractive fields assume a part in different heavenly peculiarities, including sun powered action and heavenly flares. These attractive fields can impact the way of behaving of stars and their communications with the interstellar medium.

Tyson investigates the effect of attractive fields on heavenly and cosmic cycles. He talks about what attractive movement means for star spots, sunlight based breezes, and the general way of behaving of stars.

The Grandiose Web and Enormous Scope Structure

The enormous scope design of the universe is coordinated into a huge inestimable web, with systems and world bunches framing interconnected fibers and voids. This web-like design results from the transaction of gravity, dull matter, and enormous extension.

1. Galactic Fibers and Voids:
The astronomical web comprises of cosmic fibers — long, string like designs where systems are thickly pressed — and voids, huge locales with somewhat couple of universes. The dispersion of systems along these fibers mirrors the basic gravitational powers and the impact of dim matter.

Tyson frequently alludes to the infinite web as a "enormous embroidery" that represents the huge scope association of the universe. He makes sense of how the investigation of this construction gives bits of knowledge into the powers forming the universe.

2. Galaxy Bunches and Superclusters:
World groups are monstrous convergences of systems bound together by gravity. These bunches are the biggest gravitationally bound structures known to man. Superclusters are considerably bigger developments comprising of different world groups.

Tyson talks about the meaning of universe bunches and superclusters in grasping enormous design. He features how these developments give hints about the dispersion of dim matter and the advancement of the universe.

Contextual investigation: The Recognizable Universe

To delineate the more extensive setting of infinite development, let us inspect the observable universe — the part of the universe that we can see from Earth. The perceptible universe contains billions of systems, each with its own complicated construction and history. The investigation of this huge region gives a window into the cycles of vast development and the interchange of powers forming the universe.

Neil deGrasse Tyson often involves the discernible universe as a contextual investigation to investigate inestimable peculiarities. He examines how perceptions of far off worlds and grandiose designs uncover the set of experiences and elements of the universe, offering bits of knowledge into the powers driving its advancement.

CHAPTER 7:

The Quest for Extraterrestrial Life: Perspectives and Implications

The quest for extraterrestrial life has enthralled human creative mind for quite a long time, driven by a principal interest in our spot in the universe. As we investigate the universe and see more about heavenly and cosmic cycles, whether or not we are distant from everyone else in the universe turns out to be progressively appropriate. This section looks at the logical, philosophical, and observational parts of the journey for extraterrestrial life, integrating experiences from Neil deGrasse Tyson to give an exhaustive outline of this significant subject.

The Fermi Conundrum: Grasping the Silence

One of the focal inquiries in the quest for extraterrestrial life is the Fermi Paradox — the evident inconsistency between the high likelihood of extraterrestrial human advancements existing in the universe and the absence of proof for or contact with such developments.

1. The Drake Equation:
 The Drake Condition is a device used to gauge the quantity of cutting edge developments in the world that may be fit for speaking with us. It factors in factors like the pace of star arrangement, the negligible portion of stars with planetary frameworks, and the probability of life creating on appropriate planets.

Neil deGrasse Tyson regularly examines the Drake Condition in his investigations of the Fermi Oddity. He features how this condition helps approach the quest for extraterrestrial life and highlights the vulnerabilities in assessing the quantity of likely civic establishments.

2. Possible Answers for the Fermi Paradox:
Different answers for the Fermi Catch 22 have been proposed, going from the possibility that extraterrestrial civilizations are keeping away from us, to the likelihood that exceptional developments fall to pieces prior to accomplishing interstellar correspondence, or that they exist however are just excessively far away for us to recognize.

Tyson investigates these potential arrangements, underscoring the difficulties of recognizing extraterrestrial developments and the limits of our ongoing innovation. He likewise thinks about how conceivable it is that we may not as yet be searching for the right signals or utilizing the right techniques.

The Quest for Extraterrestrial Knowledge (SETI)

The Quest for Extraterrestrial Knowledge (SETI) is a logical undertaking pointed toward recognizing signals from cutting edge extraterrestrial developments. SETI utilizes different techniques and advances to tune in for expected signals from space.

1. Radio Stargazing and Transmission Detection: SETI fundamentally utilizes radio telescopes to look for signals from different developments. The thought is that cutting-edge developments could utilize radio waves to impart, and these transmissions could be distinguished as they travel through space.

Tyson frequently examines the mechanical parts of SETI, including the utilization of huge radio exhibits and refined calculations to investigate information. He features the difficulties of recognizing expected signals from foundation commotion and the significance of proceeded with mechanical progression.

2. The Advancement Listen Initiative:
The Advancement Listen Drive is a new work to look for extraterrestrial signs, supported by confidential magnanimity. This drive means to examine the whole sky for signals across a wide scope of frequencies, utilizing probably the most progressive telescopes and hardware that anyone could hope to find.

Tyson upholds the objectives of the Advancement Listen Drive, underlining propelling comprehension we might interpret the universe and increment the possibilities identifying extraterrestrial signals potential. He examines the drive's far reaching approach and the significance of supported endeavors in the quest for extraterrestrial knowledge.

Exoplanets and the Livable Zone

The revelation of exoplanets — planets circling stars past our planetary group — has extended how we might interpret where life could exist. The idea of the livable zone, or the "Goldilocks Zone," is pivotal in deciding if exoplanets could uphold fluid water and possibly life.

1. Finding Exoplanets:
Strategies like the travel strategy, where a planet passes before its star and causes a brief diminishing, and the spiral speed technique, which estimates the star's wobble because of a planet's gravitational impact, are utilized to identify exoplanets.

Tyson frequently talks about the meaning of exoplanet revelations in the quest forever. He features how these revelations give significant data about the circumstances on different planets and their capability to help life.

2. The Livable Zone:
The livable zone is the district around a star where conditions are perfect for fluid water to

exist. Planets inside this zone are viewed as possible contender forever. In any case, factors like barometrical structure, planetary topography, and attractive fields additionally assume basic parts in livability.

Tyson investigates the idea of the livable zone and its suggestions for tracking down life. He underlines that while the tenable zone is a key variable, different circumstances should likewise be met for a planet to help life.

Extremophiles and Life's Adaptability

The investigation of extremophiles — organic entities that flourish in outrageous conditions on The planet — has extended how we might interpret the potential for life in different circumstances. Extremophiles get by in conditions like remote ocean vents, acidic lakes, and frosty districts, proposing that life could exist in comparative outrageous circumstances somewhere else in the universe.

1. Implications for Extraterrestrial Life:
The presence of extremophiles suggests that life might actually get by in brutal conditions on different planets or moons, growing the scope of where we could track down life.

Tyson talks about the ramifications of extremophiles for the quest for extraterrestrial life, featuring how these disclosures widen the extent of tenable conditions and illuminate our quest for life past Earth.

2. Astrobiology:
Astrobiology is the logical field committed to concentrating on the beginnings, development, and conveyance of life in the universe. It joins components of science, science, and planetary science to comprehend how life could emerge and get by in various conditions.

Tyson upholds the interdisciplinary methodology of astrobiology, accentuating its job in propelling comprehension we might interpret life's likely in the universe. He talks about how astrobiology incorporates discoveries from different logical disciplines to investigate the chance of life past Earth.

The Philosophical and Social Effect of Finding Extraterrestrial Life

The revelation of extraterrestrial life would have significant philosophical and social ramifications, testing how we might interpret life, knowledge, and our position in the universe.

1. Implications for Human Identity:
Finding extraterrestrial life would provoke significant inquiries concerning human personality, reason, and our position in the universe. It could reshape how we might interpret life and knowledge and our job inside the more extensive universe.

Tyson investigates the likely effect of such a disclosure on human culture, culture, and reasoning. He features the manners by which our viewpoint on presence and our relationship with the universe could change.

2. The Quest for Meaning:
The quest for extraterrestrial life isn't simply a logical undertaking yet in addition a mission for importance. It mirrors humankind's well established interest and want to grasp our position in the universe.

Tyson underlines the more extensive meaning of the quest for extraterrestrial life, talking about how it reflects central parts of human instinct and our mission for information.

CHAPTER 8:

The Future of Space Exploration: Visions and Technologies

The destiny of room examination commitments to be an outright exhilarating and notable period, put aside by degrees of progress in development, forceful missions, and new disclosures. As humanity looks past Earth, the potential for examination, colonization, and consistent jump advances creates. This part researches the possible destiny of room examination, focusing in on emerging progressions, visionary missions, and the consequences for humanity's work known to mankind. Encounters from Neil deGrasse Tyson will edify the conceivable outcomes and hardships of our future endeavors in space.

Emerging Headways in Space Exploration

Types of progress in advancement are basic to the possible destiny of room examination. From imaginative push systems to cutting brink mechanical innovation, these advances will expect a fundamental part in developing our compass and limits in space.

1. Propulsion Systems:
New drive propels are major for diminishing travel times and extending the capability of room missions. Improvements, for instance, nuclear warm propulsion and ion propulsion promise to engage speedier and more useful travel inside our planetary gathering to say the least.

Neil deGrasse Tyson every now and again analyzes the capacity of these general drive structures in changing space examination. He includes how these progressions could make missions to Mars and other distant complaints more possible and viable.

2. Robotic Exploration:

Mechanical innovation expect an essential part in space examination, from wanderers on Mars to free rocket researching distant moons. Advances in mechanized thinking and mechanical innovation will update our ability to research and sort out distant universes.

Tyson emphasizes the meaning of robotized missions in exploring conditions that are exorbitantly dangerous for individuals. He figures out how robots can aggregate significant data, direct tests, and get ready for future human missions.

3. Space Domains and Life Support Systems:

The progression of functional space domains and life sincerely strong organizations is basic for long-range missions and future space colonization. Advancements in shut circle life support, regular environmental factors plan, and resource utilization will be major for staying aware of human life in space.

Tyson analyzes the troubles and entryways related with building and staying aware of space regions. He includes consistent investigation and progressions highlighted laying out self-supporting circumstances for space explorers and future space pioneers.

Visionary Missions and Goals

The destiny of room examination consolidates an extent of forceful missions and goals highlighted expanding our presence and understanding of the universe. These missions reflect humanity's desires to examine new unsettled areas and achieve basic accomplishments.

1. Mars Exploration:
Mars is a fundamental target for future examination, with missions needed to focus on

its geology, climate, and potential for past or present life. Tries, for instance, NASA's Artemis program expect to spread out a legitimate human presence on Mars and lead coherent investigation.

Tyson habitually focuses on the importance of Mars examination in impelling understanding we could decipher planetary science and the potential for human colonization. He inspects the specific and vital challenges of sending individuals to Mars and spreading out a long presence.

2. Lunar Examination and Colonization:
The Moon fills in as a wandering stone for more significant space examination. The Artemis program plans to return individuals to the Moon, spread out a lunar base, and explore its resources. The Moon's closeness makes it an ideal region for testing developments and preparing for missions to Mars.

Tyson explores the implications of lunar examination for future space missions. He includes the intelligent entryways presented by focusing on the Moon's surface and the probable benefits of utilizing lunar resources.

3. Exploration of Outside Planets and Moons:
Missions to outside planets and their moons, similar to Jupiter's Europa and Saturn's Enceladus, are critical for getting a handle on the potential for life in our planetary gathering. Looming missions, for instance, NASA's Europa Clipper, hope to research these distant universes and journey for signs of something happening on a deeper level.

Tyson discusses the energy and troubles related with exploring the outside planets and their moons. He highlights the intelligent advantage of focusing on these chilly universes and the potential for profound disclosures.

Worldwide Composed exertion in Space Exploration

The destiny of room examination will logically rely upon overall composed exertion. The hardships and costs related with space missions are much of the time unnecessarily ideal for any single country to deal with alone, provoking associations and joint undertakings.

1. Global Space Agencies:
Workplaces like NASA, ESA (European Space Association), Roscosmos (Russian space association), and CNSA (China Public Space Association) are cooperating on various missions and adventures. These associations empower the sharing of inclination, resources, and data.

Tyson includes the benefits of worldwide coordinated effort in space examination. He inspects how joint missions and shared targets add to a more extensive cognizance of room and advance overall sensible data.

2. Commercial Spaceflight:
The rising of exclusive organizations in spaceflight, as SpaceX, Blue Start, and Boeing, has changed the location of room examination. These associations are developing new progressions, shipping off missions, and setting out open entryways for private undertakings in space.

Tyson explores the impact of business spaceflight on the destiny of examination. He highlights the occupation of exclusive organizations in driving progression, diminishing expenses, and developing permission to space.

Moral and Philosophical Considerations

As humankind undertakings further into space, moral and philosophical considerations will expect a critical part in coordinating our exercises and decisions. Requests with respect with the impact of room examination on various universes, the defending of space conditions, and the potential for planetary protection ought to be tended to.

1. Planetary Protection:
Planetary protection incorporates safeguarding other glorious bodies from debasement by Earth-starting life as well as the reverse way around. Ensuring that missions don't mull over genuineness of various universes is essential for saving intelligent accuracy and moral standards.

Tyson discusses the meaning of planetary security in staying aware of the flawlessness of consistent examination. He includes the necessity for shows and practices that thwart contamination and save the normal state of various planets and moons.

2. Human Impact and Responsibility:
The augmentation of human activities into space raises issues about our commitment to the universe and the normal impact of our presence. Thoughts consolidate the ethical repercussions of room colonization, resource use, and the protection of room conditions.

Tyson examines the greater philosophical requests associated with humankind's part in space. He complements the necessity for skilled examination and the ethical considerations related with our relationship with the universe.

The Destiny of Room Examination: Open entryways and Challenges

The destiny of room examination presents both animating entryways and gigantic troubles. As we advance in development, leave on forceful missions, and work together worldwide, we will encounter new unsettled areas and entryways for disclosure.

1. Expanding Human Presence:

The improvement of human presence into space, including the groundwork of lunar bases, Mars regions, and examination of distant universes, will be a huge point of convergence of future endeavors. These undertakings will require vanquishing particular, key, and moral hardships.

Tyson includes the normal benefits of broadening human presence in space, including sensible degrees of progress, mechanical turns of events, and the potential for long stretch perseverance. He in like manner tends to the hardships that ought to be addressed to achieve these targets.

2. Innovations and Discoveries:

The continued with examination of room will likely incite new revelations and improvements with broad repercussions. From understanding the starting points of the universe to finding new sorts of life, the destiny of room examination holds the responsibility of historic jump advances.

Tyson highlights the meaning of revenue and examination in driving sensible progression. He looks at how future disclosures in space could reshape how we could decipher the universe and our place inside it.

CHAPTER 9:

The Human Element: Psychological and Societal Impacts of Space Exploration

Space investigation isn't just a mechanical and logical undertaking yet additionally a significantly human one. As we adventure further into the universe, the mental and cultural effects on people and networks become progressively huge. This part looks at the human component of room investigation, zeroing in on the mental difficulties looked by space travelers, the cultural effect of room missions, and the more extensive ramifications for mankind as we grow our presence in space. Bits of knowledge from Neil deGrasse Tyson will assist with enlightening the individual and social components of our excursion into the universe.

Mental Difficulties of Room Travel

The mental impacts of room travel are a basic thought for long-span missions and future space colonization. Space travelers face exceptional difficulties that influence their psychological wellness and execution in space.

1. Isolation and Confinement:
Space travelers on lengthy term missions experience seclusion from Earth and imprisonment inside shuttle or natural

surroundings. These circumstances can prompt sensations of dejection, claustrophobia, and partition from loved ones.

Neil deGrasse Tyson frequently talks about the mental difficulties of disconnection and constrainment in space. He accentuates the requirement for procedures to help mental prosperity, incorporating standard correspondence with friends and family, social emotionally supportive networks, and mental advising.

2. Stress and Choice Making:
The high-stress climate of room missions can influence navigation and execution. Space travelers should oversee pressure while performing complex assignments, confronting crises, and adjusting to unanticipated difficulties.

Tyson features the significance of mental versatility and preparing for space travelers. He examines how stress the executives strategies, group building works out, and emotional wellbeing support are fundamental for guaranteeing compelling execution and prosperity during missions.

3. Adaptation to Space Environment:
The microgravity climate of room influences both physical and mental variation. Space explorers should conform to changes in their body's working, for example, muscle decay and bone thickness misfortune, as well as the mental impacts of a new and new climate.

Tyson investigates the difficulties of adjusting to microgravity and the effect on space explorers' psychological wellness. He talks about continuous examination and procedures to relieve the impacts of delayed spaceflight on mental and actual wellbeing.

Cultural Effect of Room Exploration

Space investigation has broad cultural ramifications, impacting society, instruction, and public view of science and innovation.

1. Inspiration and Education:
Space missions move and teach individuals all over the planet, starting interest in science, innovation, designing, and math (STEM) fields. The accomplishments of room investigation act as a strong inspiration for people in the future.

Tyson frequently underlines the rousing job of room investigation in advancing STEM schooling. He talks about how space missions catch the creative mind of youngsters and urge them to seek after professions in science and designing.

2. Cultural Impact:
Space investigation impacts culture by forming how we view our spot in the universe and our relationship with Earth. It can challenge existing convictions and rouse new accounts about humankind's job in the universe.

Tyson investigates the social effect of room investigation, including what it means for workmanship, writing, and media. He examines how space missions add to a more extensive comprehension of humankind's spot in the universe and motivate new social points of view.

3. International Cooperation:
Space investigation cultivates global participation and cooperation, as countries cooperate on joint missions and ventures. This participation can advance serene relations and improve worldwide logical information.

Tyson features the advantages of worldwide collaboration in space investigation. He talks about how cooperative endeavors in space add to discretionary relations and reinforce organizations between nations.

Moral and Philosophical Considerations

The extension of human exercises into space brings up moral and philosophical issues about our obligations and the effect of our activities on the universe.

1. Ethics of Space Colonization:
The likely colonization of different planets and moons brings up moral issues about our entitlement to change and occupy extraterrestrial conditions. Contemplations incorporate the conservation of expected extraterrestrial environments and the effect on people in the future.

Tyson investigates the moral contemplations of room colonization, stressing the requirement for capable investigation and regard for different universes. He examines the significance of creating moral systems to direct our activities in space.

2. Planetary Security and Preservation:
The standards of planetary security include protecting other heavenly bodies from tainting and safeguarding their regular state. Guaranteeing that our investigation exercises don't think twice about trustworthiness of these conditions is a key thought.

Tyson talks about the significance of planetary security in keeping up with logical precision and moral guidelines. He features the requirement for conventions and practices to forestall pollution and safeguard the immaculateness of extraterrestrial conditions.

Human Versatility and Adaptability

Human versatility and flexibility are essential for beating the difficulties of room investigation and guaranteeing the outcome of future missions.

1. Training and Preparation:
Space explorers go through thorough preparation and arrangement to foster the abilities and versatility required for space missions. This preparing incorporates functional preparing, specialized abilities, and mental help.

Tyson underscores the significance of thorough preparation for space travelers. He examines how preparing programs get ready people for the requests of room travel and backing their prosperity and execution.

2. Coping Mechanisms:
Creating successful survival techniques is fundamental for dealing with the mental and inner difficulties of room travel. Methodologies like care, unwinding strategies, and encouraging groups of people are significant for keeping up with psychological wellness.

Tyson investigates different survival strategies that space explorers use to oversee pressure and keep up with mental prosperity. He talks about the job of mental help and flexibility preparing in guaranteeing effective missions.

The Job of Room Investigation in Human Evolution

Space investigation can possibly affect human development by growing comprehension we might interpret life, innovation, and our spot in the universe. It might likewise impact the future improvement of human social orders and societies.

1. Technological Advancements:
The mechanical headways driven by space investigation can prompt advancements with wide applications on The planet. These progressions might change different parts of human existence, from medical services to correspondence.

Tyson examines what space investigation drives innovative advancement and its mean for on regular day to day existence. He features the manners by which space-related innovations add to progressions in numerous fields.

2. Human Point of view and Growth:
Investigating the universe can improve our point of view on human life and support development as an animal groups. It provokes us to think past our nearby worries and think about

our position in the more extensive setting of the universe.

Tyson investigates what space investigation means for human viewpoint and development. He examines how it urges us to contemplate our drawn out future and the potential for positive change.

CHAPTER 10:

Legacy and Vision: The Long-Term Impact of Space Exploration

As mankind continues to advance in space examination, the long impact of these undertakings will shape our future and leave a persevering through legacy. This segment examines the continuing on through effects of room examination on science, development, culture, and humankind overall. It examines what current and future missions will mean for individuals later on and add to the greater story of human progression. Encounters from Neil deGrasse Tyson will help with illustrating how we could decipher the practice of room examination and its significance for what's to come.

Sensible Legacy: Developing Data and Discovery

The consistent practice of room examination is separate by the critical exposures and degrees of progress that have broadened how we could decipher the universe.

1. Advances in Cosmology and Astrophysics:
Space missions have improved our understanding into the universe, from the plan of grandiose frameworks to the possibility of dull openings. Instruments like the Hubble Space Telescope and the James Webb Space Telescope

have given phenomenal encounters into the universe.

Neil deGrasse Tyson consistently discusses how space examination has advanced our understanding into significant requests in stargazing and cosmology. He highlights the meaning of these disclosures in trim how we could decipher colossal characteristics and the starting points of the universe.

2. Planetary Science and Astrobiology:
The examination of various planets and moons has yielded significant information about their sythesis, geography, and potential until the end of time. Missions to Mars, Europa, and Enceladus have given hints about the conditions that could maintain life.

Tyson researches the impact of planetary science and astrobiology on how we could decipher life in the universe. He includes how these missions add to the journey for extraterrestrial life and the examination of planetary systems.

3. Technological Innovations:
Space examination has driven mechanical advancements with wide applications in the world. Advances in materials science, mechanical innovation, and correspondence progressions have started from the hardships of room missions.

Tyson discusses the mechanical practice of room examination, including how headways made for space missions have dealt with various pieces of everyday presence. He includes the way space advancement has changed fields like prescription, transportation, and figuring.

Social Impact: Embellishment Human Inventive brain and Identity

Space examination has made a huge social difference, shaping how we view ourselves and our spot in the universe.

1. Inspiration and Imagination:
The achievements of room examination move ingenuity and imaginative psyche, influencing workmanship, composing, and standard society. Space missions and disclosures habitually go about as a foundation for exploring subjects of examination, experience, and human potential.

Tyson researches how space examination moves innovative verbalization and social records. He looks at the specific employment of room missions in quickening creative mind and imaginative brain in various sorts of media.

2. Global Perspective:
The viewpoint on Earth from space has developed a sensation of overall fortitude and interconnectedness. Photos of our planet from space highlight the delicacy of our ongoing situation and the prerequisite for overall support in keeping an eye on troubles.

Tyson includes the impact of seeing Earth from space on our perspective of overall issues. He looks at how space examination upholds a greater viewpoint on mankind's normal commitment and the prerequisite for total movement.

3. Human Character and Future Aspirations:
Space examination challenges how we could decipher human person and desires. It prompts us to contemplate our situation in the universe and our actual limit with regards to future turn of events and exposure.

Tyson examines the philosophical implications of room examination on human character and objectives. He discusses how the mission to explore the universe affects our vision for the future and our cognizance of being human.

Informational and Persuasive Legacy: Supporting Future Generations

The custom of room examination recollects its occupation for guidance and inspiration, shaping the cravings and goals of individuals later on.

1. STEM Guidance and Calling Paths:
Space examination has progressed income in science, advancement, planning, and math (STEM) fields. Informative tasks and exertion drives moved by space missions ask students to seek after livelihoods here.

Tyson highlights the meaning of room examination in developing STEM preparing. He looks at how space missions give educational entryways and spur youths to attract with science and development.

2. Public Responsibility and Outreach:
Public responsibility drives, for instance, space-themed presentations, talks, and media consideration, expect a fundamental part in exposing issues and interest in space examination. These undertakings help to develop

a greater understanding of the significance of room missions.

Tyson includes the occupation of public obligation to propelling space examination. He inspects how effort practices add to a more unmistakable excitement for space science and move public assistance for space missions.

3. Legacy Undertakings and Institutions:
Associations, for instance, exhibitions, research centers, and space workplaces save and advance the achievements of room examination. These legacy projects help to ensure that the responsibilities of room missions are seen and celebrated.

Tyson examines the occupation of legacy undertakings and establishments in saving the arrangement of encounters and achievements of room examination. He inspects how these drives add to our total memory and cognizance of room missions.

Future Dreams: Outlining the Way Forward

The inevitable destiny of room examination holds the responsibility of new disclosures and movements that will continue to shape our legacy and vision.

1. Ambitious Goals and Missions:
Future missions, similar to those to Mars, the Moon, to say the very least, will develop the practice of past achievements and prepare for new disclosures. These missions will test new developments, examine new backcountry, and expand how we could decipher the universe.

Tyson looks at the forceful goals and missions that lie ahead for space examination. He includes the potential for historic exposures and the persistent journey for data and examination.

2. Sustainability and Careful Exploration:
Ensuring the practicality of room examination and keeping an eye on the ethical thoughts of our activities are basic for the excessively long

advancement of future missions. Able examination practices will help with saving the decency of room conditions and advance moral lead.

Tyson focuses on the meaning of practicality and trustworthy examination in framing the destiny of room missions. He discusses the prerequisite for moral frameworks and practices that guide our relationship with the universe.

3. Human Legacy in Space:
The custom of human space examination will be described by our achievements, values, and responsibilities to the greater story of human headway. Our examination tries will leave a persevering through impact on science, culture, and our appreciation of the universe.

Tyson examines the greater implications of human space examination for our legacy and future desires. He looks at how our undertakings in space will shape the course of humankind's arrangement of encounters and effect individuals later on.

<u>CONCLUSION:</u>

<u>Embracing the Infinite Frontier</u>

The excursion of room investigation addresses quite possibly of humankind's most significant and aggressive undertaking. As we navigate the universe, driven by interest and the mission for information, we persistently rethink how we might interpret the universe and our place inside it. The investigation of room upgrades our logical and mechanical capacities as well as reshapes our social viewpoints and motivates people in the future.

Reflections on the Journey

Space investigation has prompted striking revelations, from the complicated subtleties of far off worlds to the potential for life on different planets. Every mission, whether it includes sending a wanderer to Mars, sending off a telescope into space, or arriving on the Moon, expands upon the tradition of those that preceded. Neil deGrasse Tyson's experiences have enlightened the meaning of these accomplishments, featuring how they extend our insight and push the limits of human potential.

The difficulties looked by space travelers, the mechanical developments brought into the world from space missions, and the social effect of these undertakings all highlight the extraordinary idea of room investigation. As we keep on investigating the limitless boondocks, we are not just looking for replies to major inquiries regarding the universe yet in addition endeavoring to more readily grasp ourselves and our position in the great embroidered artwork of presence.

The Human Element

Understanding the human component of room investigation is essential for guaranteeing the prosperity and progress of future missions. The mental and cultural effects of room travel, including disconnection, stress, and transformation, highlight the requirement for complete emotionally supportive networks for space explorers. Furthermore, the social and instructive tradition of room investigation moves and inspires individuals all over the planet, cultivating a feeling of worldwide solidarity and empowering the quest for STEM fields.

Neil deGrasse Tyson's appearance on the human parts of room investigation stress the significance of tending to these difficulties and guaranteeing that our excursion into space is basically as fulfilling and significant as could really be expected. By zeroing in on psychological wellness, public commitment, and moral contemplations, we can more readily explore the intricacies of room investigation and upgrade its advantages for all.

Heritage and Vision

The tradition of room investigation is described by the enduring effect it has on science, innovation, culture, and humankind. From the headways in cosmology and planetary science to the developments that work on our regular routines, space investigation has made a permanent imprint on human advancement. The motivation and creative mind it powers keep on molding our social stories and instructive goals.

Planning ahead, the aggressive objectives and missions not too far off vow to expand on this inheritance and impel us higher than ever. As we seek after supportability and dependable investigation, we guarantee that our undertakings are directed by moral contemplations and a promise to protecting the uprightness of room conditions.

Neil deGrasse Tyson's vision for space investigation welcomes us to embrace the endless outskirts with a feeling of marvel and obligation. The quest for information and revelation will keep on driving humankind's excursion into the universe, offering new experiences and potential open doors for development.

The Boundless Frontier

As we stand near the precarious edge of new revelations and difficulties, the boundless wilderness of room allures us to investigate further and dream greater. The excursion of room investigation is a demonstration of human creativity, persistence, and interest. It mirrors our craving to grasp the universe and to push the limits of what is conceivable.

In this pursuit, we not just look to respond to central inquiries concerning the universe yet additionally to move people in the future to proceed with the mission for information and investigation. By embracing the boundless outskirts earnestly and vision, we honor the tradition of the people who have made ready and add to a future loaded up with vast potential outcomes.

As we adventure into the obscure, let us convey forward the soul of investigation, directed by the experiences of Neil deGrasse Tyson and the aggregate insight of the individuals who have wandered into the universe before us. The excursion of room investigation is a demonstration of our persevering through journey for understanding and our steadfast craving to investigate the stars.

The investigation of room isn't simply an excursion into the universe; it is an excursion into the core of human potential and the endless conceivable outcomes that look for us among the stars.

APPENDIX:

Resources, Terminology, and Further Reading

The reference region gives extra assets and data to improve the significant text of this book. It combines a glossary of key terms, a quick overview of significant assets for additional review, and extra investigating materials that can extend how you could translate space assessment and Neil deGrasse Tyson's liabilities to the field.

Glossary of Key Terms

1. Astrobiology: The predictable assessment of the beginning, progress, and arrangement of life in the universe.

2. Cosmology: The piece of room science that spotlights on the beginning, movement, and undeniable destiny of the universe.

3. Exoplanet: A planet that circles a star outside our nearby planet bundle.

4. Microgravity: The condition wherein objects show up, evidently, to be weightless considering the especially low gravitational powers experienced in space.

5. Nuclear Warm Propulsion: A space gadget force strategy that utilizes an atomic reactor to warm a charge, conveying push.

6. Planetary Protection: Measures taken to ruin contamination of other great bodies by Earth-beginning life structures as well as the opposite strategy for getting around.

7. Space Habitat: A living climate made arrangements for human inhabitance in space, for example, a space station or lunar base.

8. Stellar Evolution: The correspondence by which a star changes throughout a drawn out

time, including its new development, lifecycle, and preposterous destiny.

9. Theoretical Physics: A piece of material science that utilizes numerical models and reflections to sort out and foresee standard qualities.

10. Voyager Probes: a few vehicle sent off by NASA in 1977 to zero in outwardly planets and occur into interstellar space.

Suggested Resources

1. NASA's Website: [NASA.gov](https://www.nasa.gov) - The power page for NASA, including news, examination, and updates on space missions and divulgences.

2. Space.com: [Space.com](https://www.space.com) - A broad point of convergence for space news, articles, and parts on stargazing and space assessment.

3. European Space Affiliation (ESA): [ESA.int](https://www.esa.int) - The power site for ESA, giving data on European space missions and examination.

4. SpaceX: [SpaceX.com](https://www.spacex.com) - The power site for SpaceX, recalling restores for business spaceflight and space missions.
5. The Planetary Society: [Planetary.org](https://www.planetary.org) - A useful connection zeroed in on space assessment and backing.

Further Reading

1. "Astrophysics for Individuals in a Rush" by Neil deGrasse Tyson - A brief and drawing in design of stargazing and the universe.
2. "The Boundless Alliance: An Extraterrestrial Point of view" through Carl Sagan - Explores the potential for extraterrestrial life and the meaning of room assessment.
3. "Cosmos" through Carl Sagan - An estimable work that looks at the universe and our place in it through a mix of reasonable experiences and philosophical reflections.
4. "The Universe Basically" by Stephen Hawking - An improvement to Selling's "A

Short History of Time," introducing a cutting edge point of view on cosmology and speculative material science.

5. "The Martian" by Andy Weir - A craftiness that gets reasonable accuracy along with stimulating portraying to portray a man's battle for constancy on Mars.

6. "Cosmic Solicitations: StarTalk's Manual for What Our personality is, how We Appeared, and Where We're Going" by Neil deGrasse Tyson - Replies to intriguing solicitations concerning the universe and our place inside it.

7. "Astrobiology: A Remarkably Short Show" by David C. Catling - An open plan of astrobiology and the excursion for life past Earth.

Extra References

- NASA's Astrobiology Institute: Examination and assets on the assessment of life in the universe.

- The Hubble Space Telescope Archives: Pictures and information from possibly of the basic telescope in space.

- James Webb Space Telescope (JWST) Updates: Data and appraisal from the cutting edge space telescope.
- Overall Space Station (ISS) Research: Subtleties on the reasonable appraisal composed on board the ISS.

ACKNOWLEDGEMENTS

The production of this book has been a cooperative exertion, drawing on the skill and experiences of various people and establishments committed to the field of room investigation. I'm profoundly thankful to the people who have added to this work and made it conceivable.

I, first and foremost, might want to stretch out my genuine thanks to Neil deGrasse Tyson. His significant information, drawing in correspondence style, and devotion to making complex logical ideas available have been a priceless wellspring of motivation. His experiences into the idea of the universe and space investigation have enormously advanced this book.

I'm likewise appreciative to the numerous researchers, designers, and specialists who keep on propelling comprehension we might interpret space. Their notable work, from the investigation of far off planets to the investigation of heavenly development, shapes the groundwork of the information examined in these pages.

An extraordinary thank you to the establishments that help space investigation and science schooling, including NASA, the European Space Organization, SpaceX, and The Planetary Society. Their missions and examination give the structure to our investigation of the universe and proposition important assets for figuring out our spot in the universe.

To my partners and counselors who looked into drafts, gave input, and offered their mastery — your commitments have been vital for the turn of events and refinement of this book. Your help and support have been incredibly valuable.

Ultimately, I might want to recognize my loved ones for their understanding, understanding, and relentless help all through the creative cycle. Your faith in the significance of this work has been a wellspring of inspiration and motivation.

This book is a demonstration of the cooperative soul of the individuals who look to investigate and figure out the universe. It is my expectation that the experiences shared inside these pages will motivate perusers to keep investigating the marvels of room and add to the continuous journey for information.